For Jaka and Oliver,
my favorite hipsters in Silicon Valley.

Life of a Silicon Valley Hipster

Written by Jasmine Jaksic
Illustrated by Tjaša Buh

On a bright and sunny day,
Theo drove back to San Fran,
High as balls on magic shrooms,
Reeking still of Burning Man.

Like a sudden lightning bolt,
An idea struck his brain,
An earth shattering startup,
Cleverest thing since blockchain.

He searched all the app stores,
Twitter, Pinterest and blogs,
With no matches to speak of,
He started "Tinder for dogs".

Wearing his polka dot shirt,
And his very skinny jeans,
He donned his red bow-tie,
Ready to realize his dreams.

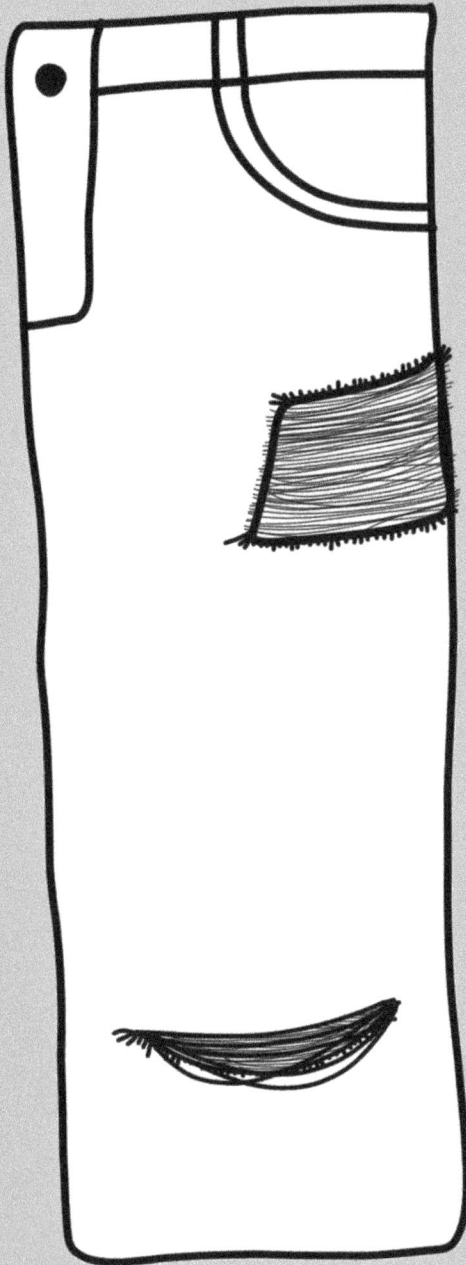

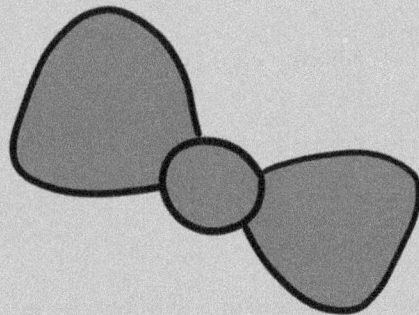

Biking through San Francisco,
Avoiding urine and poo,
He arrived at a Starbucks,
To meet an investor crew.

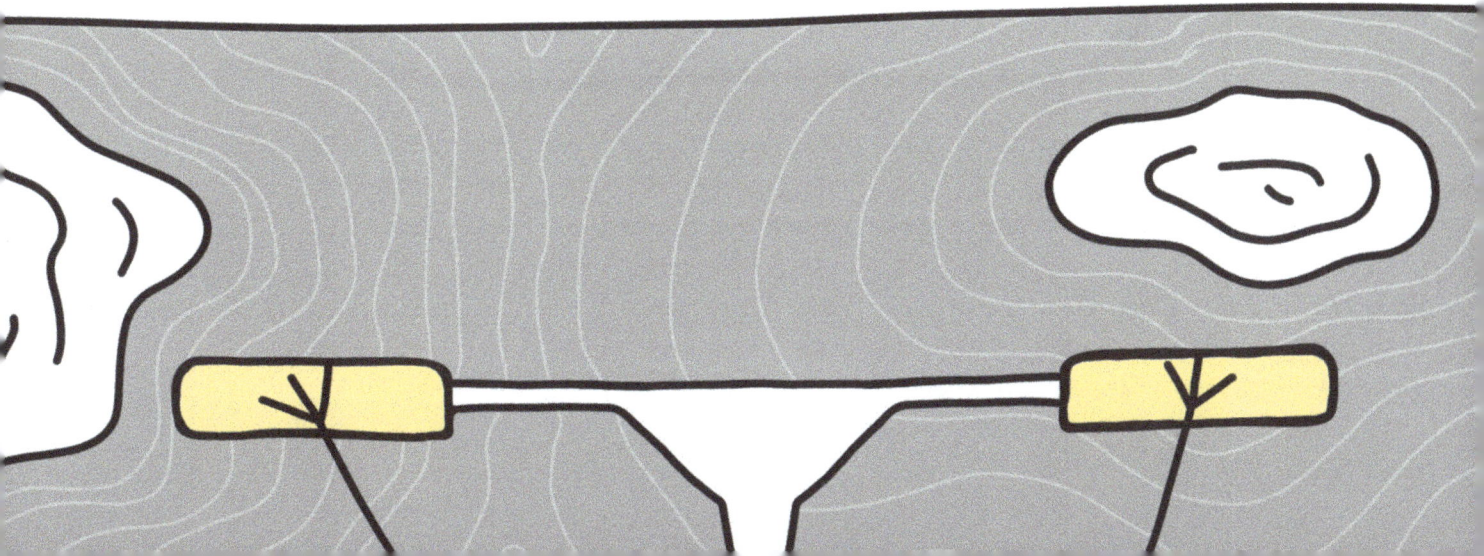

He boldly presented his plan,
It lacked a revenue path.
But he promised rapid growth,
So VC's didn't care about math.

Eager to cause disruption
In the pet dating space,
He hired two high-paid engineers,
To make the world a better place.

BUSINESS
PLAN

1. ROCKSTAR ENGINEERS ✓
2. STOCKED KITCHEN ✓
3. FOOSBALL TABLE ✓
4. HIP TECHNOLOGIES ✓
5. RAPID GROWTH
6. UNICORN STATUS

The team worked night and day,
As their equity was vesting.
They soon built a prototype,
Ready for poodle testing.

Some users liked the service,
Yet the adoption was low.
They added machine learning,
But that just made the app slow.

"Fail fast and fail often",
They did plenty of that.
Despite many iterations,
Usage metrics remained flat.

FAIL
FAIL
FAIL
FAST AND
OFTEN

Having drained all the funding,
There's only one way to win.
Theo proposed a pivot,
With a cryptocurrency spin.

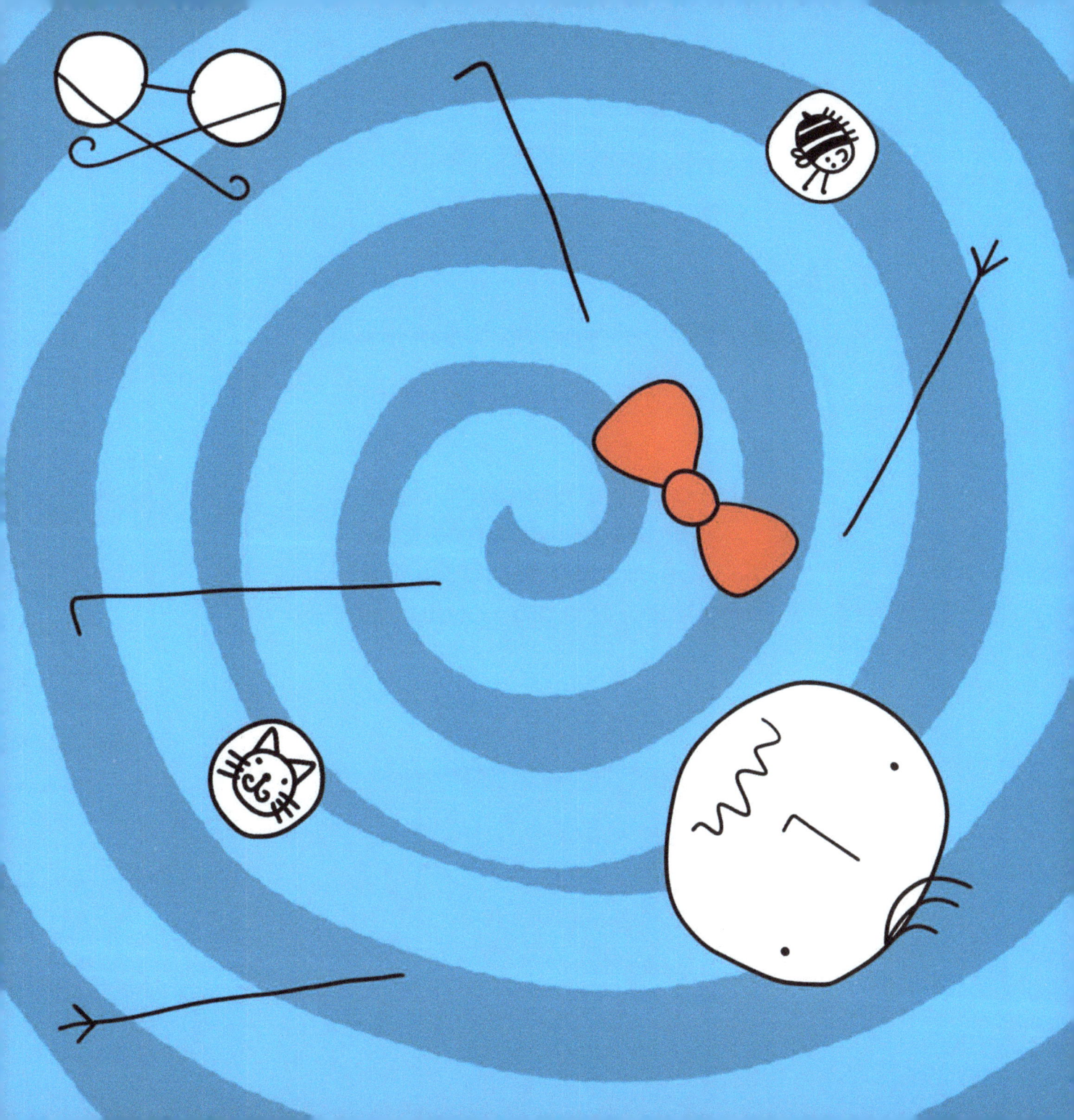

He met with investors again,
Over fried gluten free kale.
As soon as they heard "petcoin",
They knew the plan couldn't fail.

Blockchain is all the rage,
And the VCs couldn't resist.
It's a perfect solution,
For a problem that doesn't exist.

Soon they started mining coins,
And hoped to be like Ether,
With no intrinsic value,
or market stability either.

The Millennials flocked,
So did the frenzied press,
But when the money rolled in,
Soon followed the IRS.

Buried deep in late fees,
And taxes they had to pay,
They watched the bubble burst,
And the (non-intrinsic) value blow away.

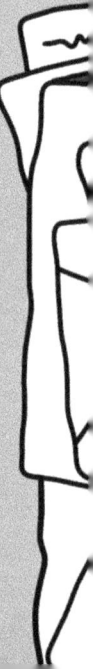

One day they were on Tech Crunch,
The next day they were gone,
Such is a startup lifecycle,
They fade as quickly as they spawn.

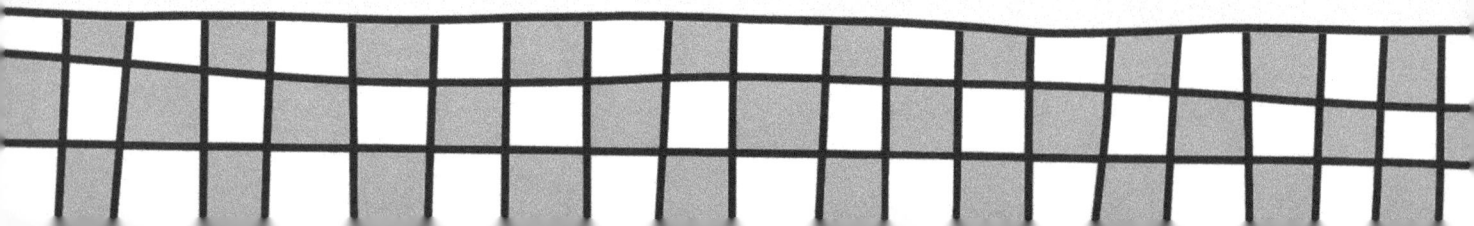

Heading back to Burning Man,
Theo knew not to fret,
Here in Silicon Valley,
Failure is just a skill set.

www.ingramcontent.com/pod-product-compliance
Lightning Source LLC
Chambersburg PA
CBHW040245100426
42811CB00011B/1157